Mercedes-Benz Museum

Max-Gerrit von Pein

STADLER

GRUSSWORT
[WELCOME

Liebe Freundinnen und Freunde
des Mercedes-Benz Museums,

am 19. Mai 2006 eröffnete das Mercedes-Benz Museum an einem traditionsreichen Ort in Stuttgart-Bad Cannstatt seine Pforten. Nur wenige Meter von hier entwickelten Gottlieb Daimler und Wilhelm Maybach zwischen 1883 und 1885 den schnell laufenden Verbrennungsmotor und schufen damit die Grundlage für unsere heutige mobile Gesellschaft – zu Lande, zu Wasser und in der Luft. Kaum eine andere Erfindung hat unsere Welt in atemberaubend kurzer Zeit so verändert.

Auf ihrem Rundgang durch das Mercedes-Benz Museum begeben sich die Besucher auf eine Zeitreise durch die Geschichte des Automobils. Begleiter dieser Reise sind 160 Fahrzeugexponate: 80 Personenwagen, 40 Nutzfahrzeuge und 40 Renn- und Rekordfahrzeuge. Aber auch die Architektur und die museale Gestaltung verkörpern, was für die Marke Mercedes-Benz charakteristisch ist: Innovationskraft, Faszination und zeitloses Design.

Im Geleitwort für den Architekturwettbewerb stand geschrieben: „Das neue Mercedes-Benz Museum soll keine Ausstellung im herkömmlichen Sinne werden, sondern ein Forum zur aktiven Kommunikation von Tradition und Herkunft der Marke Mercedes-Benz."

Wir haben eine Tradition wie kein anderer Automobilhersteller und wir sind sehr stolz darauf, denn sie ist ein wesentlicher Bestandteil unserer Identität und unserer Unternehmenskultur. Im Mercedes-Benz Museum ist sie in einer einzigartigen Art und Weise erlebbar.

Michael Bock
Geschäftsführer
Mercedes-Benz Museum GmbH

Dear Friends of the
Mercedes-Benz Museum,

On May 19, 2006, the Mercedes-Benz Museum opened its doors at a place that is steeped in tradition: in Stuttgart-Bad Cannstatt. Not far from here, Gottlieb Daimler and Wilhelm Maybach developed the high-speed internal combustion engine between 1883 and 1885, thereby laying the foundations for the mobile society as we know it today – for mobility on land, on water and in the air. There are few other inventions which have changed the world in so breathtakingly short a time.

On their tour of the Mercedes-Benz Museum, visitors set out on a journey through time and the history of the automobile – accompanied by 160 vehicle exhibits: 80 passenger cars, 40 commercial vehicles and 40 racing and record cars. But this is not all: the architecture and the exhibition design equally express what is characteristic for the Mercedes-Benz brand: innovative strength, fascination and timeless design.

In the preface to the specifications for the architectural competition, it said: "The new Mercedes-Benz Museum is to become not an exhibition in the conventional sense but a forum for the active communication of the tradition and heritage of the Mercedes-Benz brand."

We have a tradition that is unparalleled by any other manufacturer, and we are indeed very proud of it because it is a major element of our identity and corporate culture. In the Mercedes-Benz Museum, this tradition can be experienced in a unique manner.

Michael Bock
Managing Director
Mercedes-Benz Museum GmbH

INHALT
[CONTENT

TRADITION EINER MARKE
[TRADITION OF A BRAND

Mit der Erfindung des schnell laufenden Motors und des Automobils legten Gottlieb Daimler und Karl Benz in den achtziger Jahren des 19. Jahrhunderts unabhängig voneinander den Grundstein für den motorisierten Individualverkehr. In Mannheim gründete Benz im Oktober 1883 die Firma Benz & Cie., in Cannstatt entstand im November 1890 die Daimler-Motoren-Gesellschaft (DMG).

Um ihre Produkte bekannt und unverwechselbar zu machen, suchten beide Firmen nach einprägsamen Warenzeichen. Zunächst waren es die Namen der Erfinder selbst, „Benz“ und „Daimler“, die für den Ursprung und die Qualität der Motoren und Fahrzeuge bürgten. Doch während sich der Markenname von Benz & Cie. bis zur Fusion beider Unternehmen im Jahr 1926 namentlich nicht veränderte, tauchte für die Fabrikate der Daimler-Motoren-Gesellschaft (DMG) nach der Jahrhundertwende der völlig neue Markenname „Mercedes“ auf.

With the invention of the high-speed engine and the automobile, Gottlieb Daimler and Karl Benz laid the foundations for motorized private transport in the 1880s – quite independently of one another. Benz founded his company, Benz & Cie., in Mannheim in October 1883; a company named Daimler-Motoren-Gesellschaft (DMG) was established in Cannstatt in November 1890.

To make their products both known and unmistakable, the two companies looked for distinctive trademarks. Initially, the names of the inventors themselves, "Benz" and "Daimler", guaranteed the origin and the quality of engines and vehicles. While the brand name of Benz & Cie. did not change until the merger of the two companies in 1926, a completely new brand name – "Mercedes" – was chosen for the products of Daimler-Motoren-Gesellschaft (DMG) after the turn of the century.

Mercedes ist ein spanischer Frauenname und bedeutet „Gnade". So hieß auch die 1889 geborene Tochter des in Baden bei Wien und in Nizza lebenden österreichischen Geschäftsmanns Emil Jellinek. Er war ein moderner und sportinteressierter Mann und ein begeisterter Anhänger des Automobils und des technischen Fortschritts und er war überzeugt, dass das Automobil die Zukunft verändern würde.
Von der Daimler-Motoren-Gesellschaft forderte Jellinek immer stärkere und schnellere Fahrzeuge und startete damit ab 1899 auch bei Rennveranstaltungen – allen voran der Woche von Nizza. Bei diesen Rennen trat er gewöhnlich unter seinem Pseudonym „Mercédès" an.
Anfang April 1900 trafen die DMG und Jellinek eine Vereinbarung über den Vertrieb von Daimler-Wagen und -Motoren. Am 22. Dezember 1900 lieferte die DMG den ersten mit einem neuen Motor ausgerüsteten Wagen an Jellinek, einen 35 PS-Rennwagen. Dieser erste „Mercedes" sorgte zu Beginn des Jahrhunderts für Furore. Mit seinem niedrigen Schwerpunkt, Pressstahlrahmen, dem leichten und leistungsstarken Motor und dem Bienenwabenkühler gilt er heute als erstes modernes Automobil.

Die Woche von Nizza im März 1901, bei der die eingesetzten Mercedes-Wagen in praktisch allen Disziplinen unschlagbar waren, verhalf Jellinek und dem Mercedes zu außergewöhnlicher Publizität. Am 23. Juni 1902 wurde der Name „Mercedes" als Warenzeichen angemeldet und am 26. September gesetzlich geschützt. Emil Jellinek erhielt im Juni 1903 die Erlaubnis, sich fortan Jellinek-Mercedes zu nennen.

Mercedes is a Spanish name and means "grace". It was the name given to the daughter born in 1889 to Austrian businessman Emil Jellinek who lived in Baden near Vienna and in Nice. He was a modern man, interested in sports and enthusiastically devoted to technical progress and the automobile, and he was convinced that the automobile would change the future.
Jellinek demanded ever more powerful and faster cars from Daimler-Motoren-Gesellschaft and from 1899 entered these in racing events – notably in the Nice Race Week. He usually raced under the pseudonym "Mercédès".

In early April 1900, DMG and Jellinek concluded an agreement on the marketing of Daimler cars and engines. On December 22, 1900, DMG supplied the first car powered by a new engine to Jellinek, a 35 hp racing car. This car – the first "Mercedes" – caused quite a stir at the beginning of the 20th century. With its low center of gravity, pressed-steel frame, lightweight and powerful engine and honeycomb radiator, it is today considered to have been the first modern automobile.

The Nice Race Week in March 1901, during which the Mercedes cars entered proved to be unbeatable in virtually all disciplines, created extraordinary public awareness of Jellinek and Mercedes. On June 23, 1902, the name "Mercedes" was registered as a brand name and received legal protection on September 26. In June 1903, Emil Jellinek obtained permission to call himself Jellinek-Mercedes from then on.

Mercedes 1909 | Benz 1909 | Mercedes 1916 | Mercedes-Benz 1926 | Mercedes-Benz 1933 | Mercedes-Benz 1989

Die DGM verfügte zwar über einen erfolgreichen Namen, aber es fehlte ein charakteristisches Warenzeichen. Gottlieb Daimlers Söhne, Paul und Adolf, erinnerten sich, dass ihr Vater einst einen Stern als Symbol nutzte. Daimler hatte zu Beginn seiner Tätigkeit in Deutz sein Wohnhaus auf einer Ansicht mit einem Stern markiert und angedeutet, dass dieser Stern einmal segensreich über seinem Werk aufgehen werde. Der Vorstand der DGM griff diesen Vorschlag auf und meldete in Juni 1909 einen Dreizack- und einen Vierzackstern als Warenzeichen an. Beide wurden geschützt, doch nur der Dreizackstern war ab 1910 auf dem Kühler in plastisch geprägter Form zu sehen. Der Dreizackstern sollte auch als Symbol der universellen Motorisierung „zu Lande, zu Wasser und in der Luft“ stehen.
Mit dem Zusammenschluss beider Firmen im Jahr 1926 fusionierten auch die Schutzmarken „Mercedes“ und „Benz“ zu „Mercedes-Benz“.

DMG had a successful name, but the company was still lacking a characteristic trademark. Gottlieb Daimler's sons Paul and Adolf remembered that their father had once used a star as a symbol. Earlier in his career, when working in Deutz, Daimler had marked his house with a star on a postcard, writing that this star would once rise above his life's work for the benefit of his family. The Board of Management of DGM adopted this proposal and in June 1909 registered a three-pointed star and a four-pointed star as trademarks. Both designs obtained legal protection but only the three-pointed star was used in a three-dimensional design on the radiator from 1910. The three-pointed star was meant as a symbol for universal motorization "on land, on water and in the air". With the merger of the two companies in 1926, their brand names "Mercedes" and "Benz" were joined into "Mercedes-Benz".

GOTTLIEB DAIMLER

Gottlieb Daimler wurde am 17. März 1834 in Schorndorf bei Stuttgart geboren. 1848 bis 1852 absolvierte er eine Ausbildung zum Büchsenmacher, danach ein Maschinenbau-Studium an der Polytechnischen Schule in Stuttgart. 1865 traf er Wilhelm Maybach und es begann eine Zusammenarbeit, die beide Männer bis an ihr Lebensende vereinte. Von 1872 bis 1882 war Daimler Technischer Direktor und Vorstandsmitglied der Gasmotoren-Fabrik in Deutz. Mit Maybach entwickelte er 1883 den ersten schnell laufenden Viertaktmotor und meldete später die „Standuhr“ 1885 zum Patent an. Sie trieb den Reitwagen an, das erste Motorrad der Welt. 1886 entstanden das erste vierrädrige Automobil und das erste Motorboot der Welt. 1890 gründete er die Daimler-Motoren-Gesellschaft (DGM). Am 6. März 1900 starb Gottlieb Daimler in Cannstatt. *Gottlieb Daimler was born in Schorndorf near Stuttgart on March 17, 1834. Between 1848 and 1852, he trained as a gunsmith before studying mechanical engineering at the Polytechnic in Stuttgart. In 1865 he met Wilhelm Maybach – from then on, the two men worked together until the end of their lives. From 1872 until 1882, Daimler worked as Technical Director and member of the Board of Management of Gasmotoren-Fabrik (Gas Engine Factory) in Deutz. Together with Maybach, he developed the first high-speed four-stroke engine in 1883 and filed a patent for this engine, popularly known as the Grandfather Clock, in 1885. This engine powered the riding car, the world's first motorcycle. In 1886, they built the first four-wheeled automobile and the world's first motorboat. Gottlieb Daimler founded Daimler-Motoren-Gesellschaft (DGM) in 1890. He died in Cannstatt on March 6, 1900.*

WILHELM MAYBACH

Wilhelm Maybach wurde am 9. Februar 1846 in Heilbronn geboren. 1856 begann er – als Vollwaise – eine Ausbildung in der Maschinenfabrik des „Bruderhauses“ in Reutlingen und traf dort 1865 mit Gottlieb Daimler zusammen. Nach seiner Anstellung bei der Gasmotorenfabrik Deutz im Jahr 1872, bei der er bis zum Chefkonstrukteur aufstieg, kündigte Maybach im Jahr 1882 und schloss einen Vertrag mit Gottlieb Daimler. 1890 wurde er zum technischen Direktor der Daimler-Motoren-Gesellschaft (DGM) ernannt, verließ die Firma für kurze Zeit, trat wieder ein und wurde Vorstandsmitglied der DGM. 1900 entwickelte Maybach den ersten Mercedes und schied 1907 aus der DGM aus. Zwei Jahre später gründete er die Luftfahrzeug-Motorenbau GmbH, später Maybach Motorenbau GmbH. Am 29. Dezember 1929 verstarb Wilhelm Maybach in Bad Cannstatt. *Wilhelm Maybach was born in Heilbronn on February 9, 1846. Orphaned at an early age, he started training in the mechanical engineering company of "Bruderhaus" in Reutlingen in 1856. This was where he met Gottlieb Daimler in 1865. In 1872, he started working for Gasmotorenfabrik Deutz (Gas Engine Factory) where he rose to the position of Chief Engineer. Maybach left the company in 1882 and concluded a contract with Gottlieb Daimler. In 1890, he was appointed Technical Director of Daimler-Motoren-Gesellschaft (DGM). He left the company only to re-join it after a short period and become a member of the Board of Management of DGM. In 1900, Maybach developed the first Mercedes. He left DMG for good in 1907 and, two years later, founded Luftfahrzeug-Motorenbau GmbH (Aero Engine Company), renamed Maybach Motorenbau GmbH (Engine Company) at a later stage. Wilhelm Maybach died in Bad Cannstatt on Dec. 29, 1929.*

Gottlieb Daimler

Wilhelm Maybach

Karl Benz

KARL BENZ

Karl Benz wurde am 25. November 1844 in Karlsruhe geboren. Von 1860 bis 1864 absolvierte er ein Studium an der Polytechnischen Schule in Karlsruhe. 1883 entwickelte Benz seinen ersten Zweitaktmotor und gründete die Benz & Cie., Rheinische Gasmotorenfabrik Mannheim. Am 29. Januar 1886 erfolgte die Patentanmeldung des dreirädrigen „Fahrzeugs mit Gasmotorenbetrieb". Mit Konstruktionen wie dem Benz „Victoria" und dem „Velo" prägte Karl Benz die Automobilgeschichte. 1903 schied er aus der Geschäftsführung der Benz & Cie. aus, blieb aber bis 1929 Aufsichtsratsmitglied (ab 1926 „Daimler-Benz AG"). In Ladenburg gründete er 1906 die Firma „C. Benz Söhne", am 4. April 1929 verstarb Karl Benz in Ladenburg.

Karl Benz was born in Karlsruhe on November 25, 1844. Between 1860 and 1864, he studied mechanical engineering at the Polytechnic in Karlsruhe. In 1883, Benz developed his first two-stroke engine and founded Benz & Cie., Rheinische Gasmotorenfabrik Mannheim. On January 29, 1886, he filed a patent for his three-wheeled "vehicle with gas engine drive". With designs such as the Benz Victoria and the Velo, Karl Benz wrote automotive history. In 1903, he retired from the management of Benz & Cie. but remained a member of the Supervisory Board until 1929 (Daimler-Benz AG from 1926). In Ladenburg, he founded Messrs. C. Benz Söhne in 1906. Karl Benz died in Ladenburg on April 4, 1929.

DAS PROJEKT
[THE PROJECT

Bei Mercedes-Benz hat es Tradition, Geschichte zu schreiben. Bereits 1923 wurde im Werk Untertürkheim das erste Museum eingerichtet und es wurden Exponate gezeigt, die heute zu den frühsten und wertvollsten Kulturgütern der automobilen Geschichte zählen. Nach der Wiedereröffnung des Museums im Jahr 1951 setzte Mercedes-Benz 1961 einen weiteren Meilenstein seiner Museumsgeschichte. Zum Jubiläum „75 Jahre Motorisierung des Verkehrs" wurde in Untertürkheim ein neues Museum geschaffen, das zahlreiche Exponate aus acht Jahrzehnten Automobilgeschichte präsentierte. Anlässlich des Jubiläums „100 Jahre Automobil" wurde das Museum neu konzipiert und 1986 feierlich eröffnet.

Mercedes-Benz has been writing history by tradition. As early as 1923, the first museum was set up on the Untertürkheim plant premises, with exhibits which today rank among the earliest and most valuable cultural assets of automotive history. The museum reopened in 1951. In 1961, Mercedes-Benz set another milestone in its museum history: In time for the 75th anniversary of motorized road transport, a new museum was built in Untertürkheim to present numerous exhibits from eight decades of automotive history. On the occasion of the 100th anniversary of the automobile, the museum was redesigned and inaugurated in 1986.

Das Mercedes-Benz
Museum im Jahr 1938.

*The Mercedes-Benz
Museum in 1938.*

Nach dem Umzug 1961
bestand die Möglichkeit
deutlich mehr Fahrzeuge
auszustellen.

Bis zu seiner Schließung besuchten mehr als 10 Millionen Menschen das Mercedes-Benz Museum. Es ist somit das meistbesuchte Firmen- und Markenmuseum auf der ganzen Welt.

By the time the Mercedes-Benz Museum was closed, over ten million people had visited it. It was thus the corporate and brand museum with the largest number of visitors in the world.

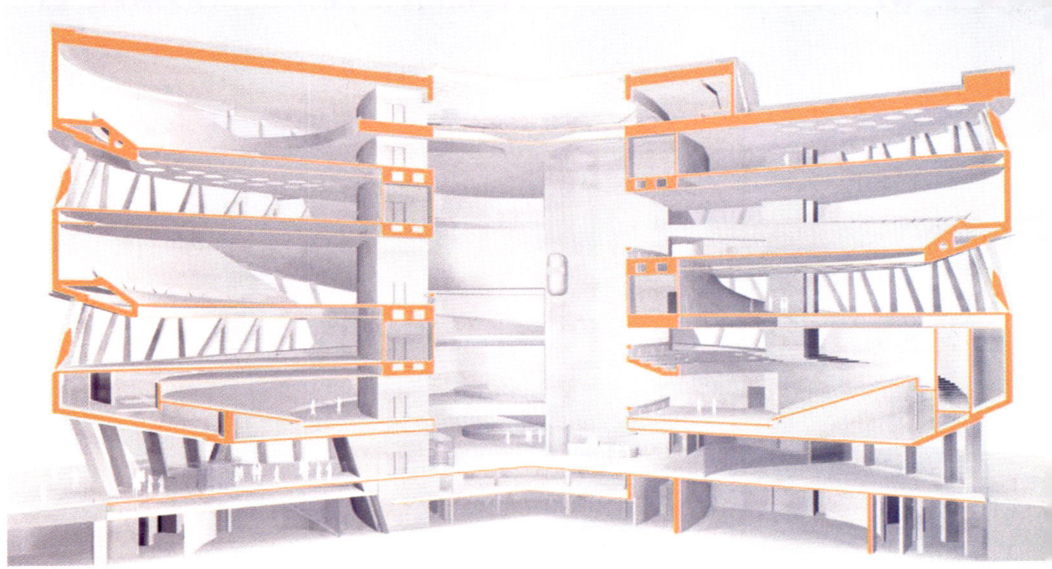

Ein Schnitt durch das neue Mercedes-Benz Museum verdeutlicht die beiden separaten Rundgänge im Gebäude.

The cut-away view of the new Mercedes-Benz Museum shows the two separate tours through the building.

Bereits seit Ende der 90er Jahre wurde intensiv an einer Neuausrichtung des Mercedes-Benz Museums gearbeitet. Im Jahr 2000 fiel die Entscheidung für den geschichtsträchtigen Standort, direkt vor den Toren des Stammwerks Stuttgart-Untertürkheim. Von 2001 bis Januar 2002 fand ein Architekturwettbewerb statt, im März 2003 wurde mit dem Bau des neuen Museums begonnen.

Am 17. September 2003 wurde der Grundstein gelegt, im März 2005 fand das Richtfest statt und am 19. Mai 2006, nach gerade einmal zweieinhalb Jahren Bauzeit, wurde das Mercedes-Benz Museum feierlich eingeweiht.

Das „Projekt Mercedes-Benz Museum" wurde zu einem weltweit einmaligen Bauvorhaben: Es ist ein Bauwerk mit anspruchsvollstem architektonischen Konzept, prototypischer Konstruktion und einem komplett neuen Museumskonzept. Es wurde innerhalb von nur drei Jahren geplant, realisiert und in Betrieb genommen.

As early as the late 1990s, a reorientation of the Mercedes-Benz Museum became the subject of intensive concept development work. In 2000, a decision was made to build a new museum at a place steeped in tradition, right outside the gates of the traditional plant in Stuttgart-Untertürkheim. An architectural design competition was staged from 2001 until January 2002, and construction of the new museum began in March 2003.

The foundation stone was laid on September 17, 2003, and the topping-out ceremony was celebrated in March 2005. Finally, on May 19, 2006, after a construction period of just two-and-a-half years, the new Mercedes-Benz Museum was inaugurated.

The Mercedes-Benz Museum became a globally unique construction project. It is a building with an extremely sophisticated architectural concept, a prototypical design and a completely new museum concept. Within a period of just three years, the museum was planned, built and put into operation.

◀ Nach der Grundsteinlegung am 17. September 2003 beginnen zuerst die Kernwände des Atriums zu wachsen.

▶ Das Projekt „Neues Mercedes-Benz Museum“ entwickelt sich zu einem der ambitioniertesten Vorhaben der Baugeschichte.

◀ *After the foundation stone was laid on September 17, 2003, construction began with the core walls of the atrium.*

▶ *The "New Mercedes-Benz Museum" project developed into one of the most ambitious undertakings in construction history.*

8
Ulm
Esslingen
S-Wangen
Großmarkt
81
Vaihingen/Enz
S-Zentrum
ZÜBLIN
LIEBHERR

Bauen im Grenzbereich des Machbaren: 16 Stunden pro Tag, sechs Tage die Woche, ein Ziel: Eröffnung im Mai 2006.

Construction near the limits of what is feasible: Sixteen hours a day, six days a week, one goal - the inauguration in May 2006.

Der Twist: Eine prototypische Konstruktion, die im Mercedes-Benz Museum erstmals gebaut wurde.

The twist: A prototypical construction, realized for the first time in the Mercedes-Benz Museum.

◄ Noch vor dem Einbringen der Exponate wurde im Rohbau des Museums der Oldtimer-Kalender 2006 produziert.

► Das Mercedes-Benz Museum bei Nacht.

◄ *The photos for the 2006 Classic Car Calendar were taken in the skeleton building, before the exhibits moved in.*

► *The Mercedes-Benz Museum by night.*

DIE ERÖFFNUNG
[THE INAUGURATION

Am 19. Mai 2006 wurde das Mercedes-Benz Museum in Anwesenheit des Ministerpräsidenten des Landes Baden-Württemberg, Günther H. Oettinger, und des Vorstandsvorsitzenden der DaimlerChrysler AG, Dr. Dieter Zetsche, im Beisein von rund 400 Gästen aus Politik, Wirtschaft und Gesellschaft feierlich eröffnet. Live zugeschaltet war die Bundeskanzlerin, Dr. Angela Merkel. An der abendlichen Gala nahmen rund 2.000 Gäste aus aller Welt teil und erwiesen dem Mercedes-Benz Museum die Ehre.

Am folgenden Wochenende, an den „Tagen der offenen Tür“, kamen fast 30.000 Menschen von nah und fern, um zu den ersten begeisterten Besuchern des Mercedes-Benz Museums zu gehören.

On May 19, 2006, the Mercedes-Benz Museum was inaugurated in the presence of Günther H. Oettinger, Minister President of Baden-Württemberg, Dr. Dieter Zetsche, chairman of the Board of Management of DaimlerChrysler AG, and some 400 guests from politics, industry and society. Federal Chancellor Dr. Angela Merkel sent a message in a live broadcast. The evening gala was attended by some 2,000 guests from all over the world, paying tribute to the Mercedes-Benz Museum.

On the following Open Days weekend, almost 30,000 people came from near and far – the first enthusiastic visitors to the Mercedes-Benz Museum.

Burkhard Wagner, Direktor der Niederlassung Stuttgart *[head of the Stuttgart company-owned sales and service outlet* Prof. Ben van Berkel, UN studio, Prof. HG Merz, Büro Merz, Max-Gerrit von Pein, Director DaimlerChrysler Heritage, und Prof. Christoph Kohlbecker, Büro Kohlbecker (v.l.n.r.).

Dr. Dieter Zetsche, Vorstandsvorsitzender von DaimlerChrysler, *[chairman of the Board of Management of DaimlerChrysler* Günther H. Oettinger, Ministerpräsident *[Minister President* und Dr. Wolfgang Schuster, Oberbürgermeister von Stuttgart *[Lord Mayor of Stuttgart* am Mercedes Simplex von 1902.

Ministerpräsident *[Minister President* Günther H. Oettinger und Dr. Dieter Zetsche, bei der symbolischen Eröffnung. *[during the inauguration ceremony.*

DIE ARCHITEKTUR
[THE ARCHITECTURE

Prof. Ben van Berkel

Ben van Berkel und Caroline Bos gründeten 1988 in Amsterdam das UNStudio, das auf Architektur, Stadtentwicklung und Verbesserungen der Infrastruktur spezialisiert ist. UN bedeutet United Network. Die Arbeit innerhalb des Netzwerkes ist zeitgemäß organisiert: Der Design-Prozess verläuft integriert ohne Hierarchien. Dabei helfen modernste Kommunikationstechnologien, die den höchstmöglichen, gegenseitigen Austausch gewährleisten. Das Büro hat seit seiner Gründung stets international gearbeitet und eine große Anzahl von Projekten realisiert, von öffentlichen Gebäuden und Infrastrukturmaßnahmen über Bürobauten bis zu Masterplänen für die Stadtentwicklung.

In 1988, Ben van Berkel and Caroline Bos founded the UNStudio in Amsterdam, specializing in architecture, urban planning and infrastructure development. UN stands for United Network. Work within this network is organized in an up-to-date manner: the design process is entirely integrated, doing without any hierarchy. This approach is supported by the latest communication technologies which ensure the most efficient mutual exchange. Ever since its foundation, the studio has been working throughout the world and realized a large number of projects, ranging from public buildings and infrastructural measures through to office buildings and master plans for urban development.

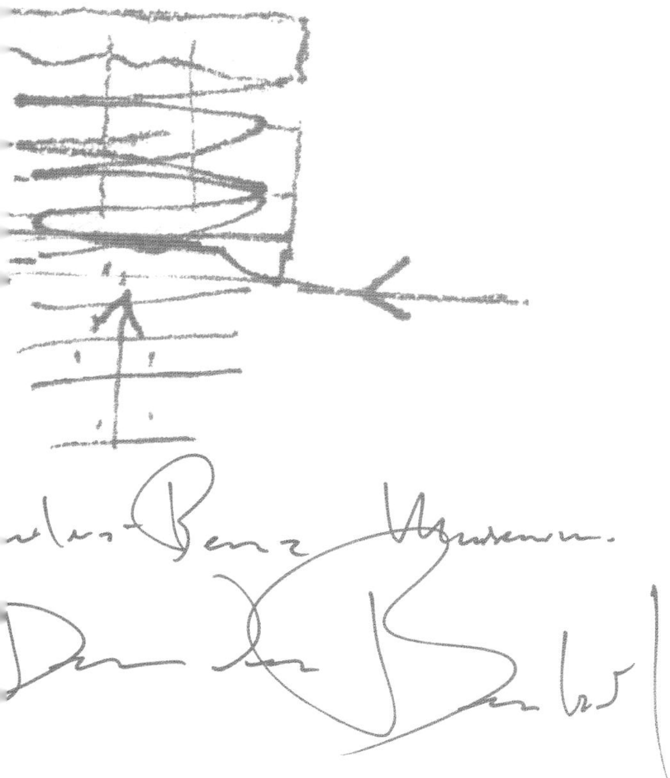

Die Architektur des Mercedes-Benz Museums geht von einem symmetrischen, dreiblättrigen Grundriss aus und die Geometrie des Gebäudes mit ihrer inneren Wegführung erinnert nicht von ungefähr an Lösungen aus dem Straßenbau: Das Gebäude korrespondiert mit dem nahe gelegenen Anschluss der Bundesstraße B14 an die Bundesstraße B10. Die Wege verschlingen sich auf neun Ebenen entlang einer Zeitachse von der Erfindung des Automobils bis in die Gegenwart. Darüber hinaus kann der Besucher einen Blick in die Zukunft des Automobils werfen.

Während der chronologische Rundgang durch die Mythos-Räume auf einer langen, stufenlosen Rampe durch die Epochen der Automobilgeschichte führt, sind die Collectionsräume durch eine zweite Spirale über schmalere Treppen an der Außenseite des Gebäudes miteinander verbunden. Auf jeder Ebene gibt es Querverbindungen, die es dem Besucher erlauben, zwischen Mythos und Collection zu wechseln.

Mythos- und Collectionsrundgang unterscheiden sich in ihrer gegensätzlichen Orientierung

The architecture of the Mercedes-Benz Museum is based on a symmetrical, three-leaved ground plan, and it is no coincidence that the building's geometry and the routes inside it are reminiscent of road construction layouts – the building corresponds with the near-by connection of highway B14 to highway B10. The routes intertwine at nine levels along a time axis, from the invention of the automobile through to the present. In addition, the visitor is able to cast a glance into the future of the automobile.

While the chronological route of the Legend rooms runs along a long, step-less ramp through the eras of automotive history, the Collection rooms are connected with each other by a second spiral with narrower staircases in the outer circumference of the building. All levels are connected with each other, permitting visitors to change from Legend to Collection and back.

The Legend and Collection tours differ in their fundamental orientation towards the inside and the outside, respectively. The visitor experiences the walk through the Legend rooms of brand history as if separated from the outside world,

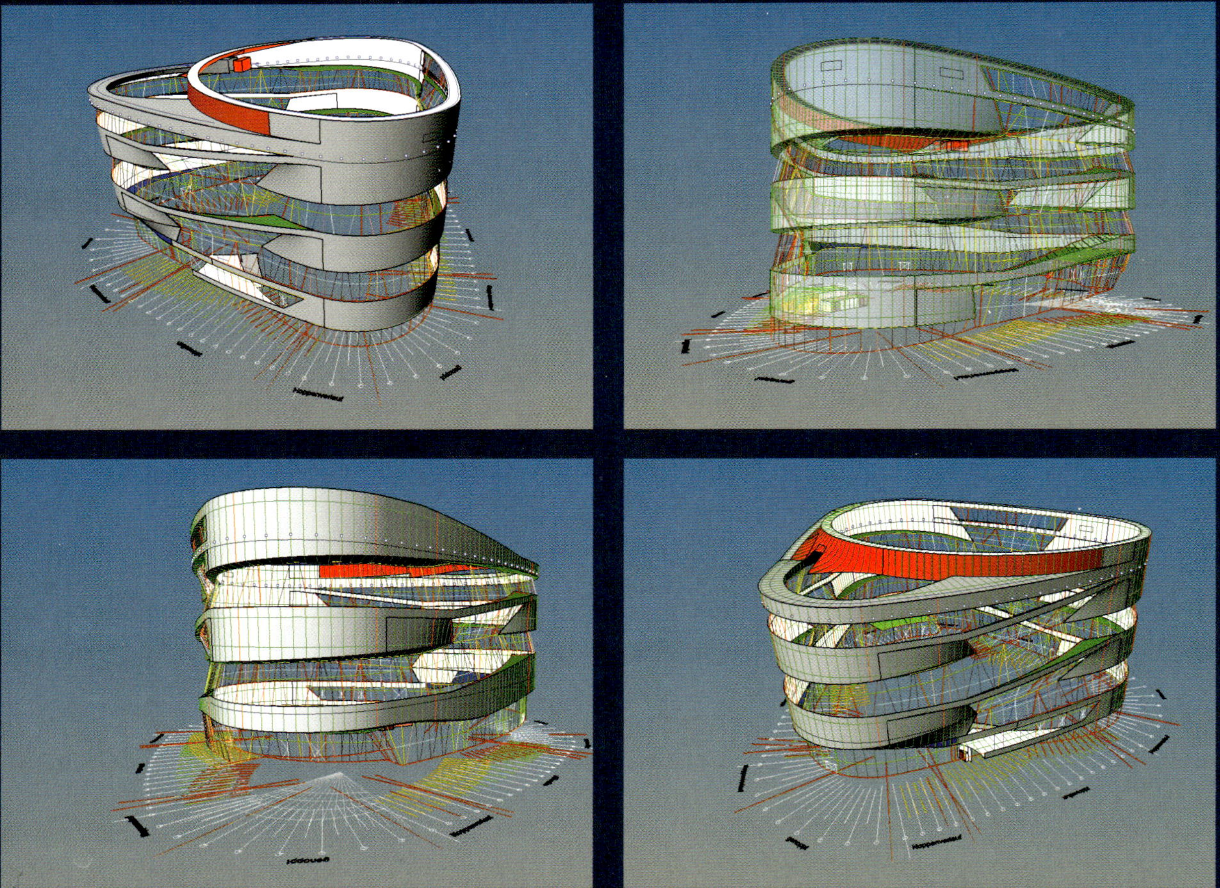

Computer-Modelle des Mercedes-Benz Museums in der Planungsphase.

Computer models of the Mercedes-Benz Museum during the planning stage

nach innen und nach außen. Den Gang durch den Mythos der Markengeschichte erlebt der Besucher, fast wie im Kino, von der Außenwelt getrennt. Die weniger hoch ausgeführten Collectionsräume dagegen sind auf der ganzen Außenseite zu einem lichten Panoramablick auf die Landschaft geöffnet.

Im gesamten Gebäude gibt es so gut wie keine rechten Winkel. Alle Wände und Decken, Rampen und Stützen sind gewölbt oder in sich gedreht und gehen in sanften, fließenden Formen ineinander über. Mehr als 110.000 Tonnen Beton und Stahl sind im Mercedes-Benz Museum verbaut, das auf 4.800 Quadratmeter Grundfläche eine Höhe von 47,5 Metern erreicht und damit einen umbauten Raum von 210.000 Kubikmeter umschließt.

almost as if in a movie theater. The light-flooded Collection rooms, by contrast, are less high and open towards the outside to provide a panoramic view of the surrounding landscape.

There are virtually no right angles throughout the building. All the walls and ceilings, ramps and supports are either curved or twisted and merge into each other in smooth, flowing movements. In the Mercedes-Benz Museum, more than 110,000 tons of concrete and steel were used to create a building with a base area of 4,800 square meters, a height of 47.5 meters and a roofed space of 210,000 cubic meters.

Die Doppelhelix mit den sieben Mythosräumen und den fünf Collectionsräumen.

The double helix with seven Legend rooms and five Collection rooms

Das Atrium hat eine lichte Höhe von 42 Metern. Drei Aufzüge befordern die Besucher nicht nur in die oberste Ebene, sondern nehmen sie mit auf eine Zeitreise aus der Gegenwart ins Jahr 1885.

The atrium is 42 meters high. Three lifts take visitors not only to the uppermost level but also on a journey through time, from the present to the year 1885.

1971 VERSCHMUTZUNGSARME RÜCKLEUCHTEN
1971 VIERSPEICHEN-SICHERHEITS
1981 AIRBAG
1979 GABELTRÄGERSTRUKTUR
Colonia

Gekrümmte Wände und anspruchsvolle Linienführungen dominieren die Architektur im Inneren des Mercedes-Benz Museums. Keine der 1.800 Fensterscheiben im Gebäude gleicht der anderen.

Curved walls and bold lines dominate the architecture inside the Mercedes-Benz Museum. Not a single one of the 1,800 window panes in the building is identical with another

Von außen stehen einzigartige Formensprachen und Rundungen im Vordergrund. Die Architektur geht über in eine Open-Air-Arena mit rund 500 Sitzplätzen, die für unterschiedlichste Veranstaltungen genutzt wird.

From the outside, the museum impresses with its unique form language and sweeping curvatures. Adjacent to the building lies an open-air arena with some 500 seats for a diversified range of events

DATEN UND FAKTEN
[FACTS AND FIGURES

Architekt [*Architect* UNStudio van Berkel & Bos, Amsterdam
Museumsgestaltung [*Museum design* Prof. HG Merz, Stuttgart
Generalübernehmer [*General contractor* DaimlerChrysler Immobilien GmbH, Berlin

Grundsteinlegung [*Laying of foundation stone* 17. September 2003
Richtfest [*Topping-out ceremony* 3. März 2005
Eröffnung [*Inauguration* 19. Mai 2006

Gebäudehöhe [*Height of building* 47,5 Meter [*meters*
Gebäudevolumen [*Roofed space* 210.000 Kubikmeter [*cubic meters*
Grundfläche [*Base area* 4.800 m^2 [*square meters*
Anzahl der Ebenen [*Number of levels* 9
Ausstellungsfläche [*Exhibition area* 16.500 m^2 [*square meters*
_ Mythosfläche [*Legend rooms* 9.100 m^2 [*square meters*
_ Collectionsfläche [*Collection rooms* 5.300 m^2 [*square meters*
_ Faszination Technik [*Fascination of Technology* 1.300 m^2 [*square meters*
Clubbereich [*Club area* 700 m^2 [*square meters*

Gezeichnete Pläne [*No. of plans drawn up* 38.000 Stück
Kabel im Beton [*Cables routed in concrete* 630.000 m [*meters*
Verträge mit Unternehmen und Ingenieurfirmen
[*No. of contracts with companies and consulting engineers* 246
Lampen im Gebäude [*No. of lamps in the building* 12.000
Polyethylenrohre [*Polyethylene pipes* 100.000 m [*meters*
Gesamtgewicht [*Total weight* 110.000 Tonnen
Fundament [*Foundation* 850 Stahlbetonpfeiler [*reinforced-concrete pillars*
Fassade [*Façade* 14.000 m^2 [*square meters*
Fensterscheiben [*No. of window panes* 1.800

Lage [*Location* 48,76 Grad nördlicher Breite 9,18 Grad östlicher Länge, 359 Meter über dem Meeresspiegel [*48.76 degrees north, 9.18 degrees east, 359 meters above sea level*

Museum

DIE AUSSTELLUNG
[THE EXHIBITION

Prof. HG Merz

HG Merz ist ein Architekt, dessen Arbeiten einen Bezug zur Geschichte haben. Seine Spezialgebiete sind die Museumsplanung und das Bauen im Bestand. Mit Mercedes-Benz ist Merz bereits seit 20 Jahren verbunden, seit er an der Neukonzeption des alten Mercedes-Benz Museums 1985/1986 beteiligt war. Großes Renommee erwarb sich HG Merz 1993 mit dem Umbau der Alten Nationalgalerie auf der Museumsinsel im Zentrum Berlins. Zum Thema Unternehmens- und Technikgeschichte erregte das zwischen 1994 und 1996 von ihm konzipierte Zeppelin-Museum in Friedrichshafen großes Aufsehen.

HG Merz is an architect whose work revolves around history. He specializes in museum planning and construction projects involving existing buildings. Merz has been associated with Mercedes-Benz for as long as 20 years – his first project for the brand had been the redesign of the old Mercedes-Benz Museum in 1985/1986. In 1993, HG Merz made a name for himself with the conversion of the Old National Gallery on the Museum Island in the center of Berlin. The concept he developed for the Zeppelin Museum in Friedrichshafen between 1994 and 1996 – a contribution in the field of corporate and engineering history – caused quite a sensation.

So ist HG Merz nicht nur für die Präsentation der Sammlung und der Unternehmensgeschichte im Mercedes-Benz Museum verantwortlich, auch an der Konzeption hat er von Anfang an mitgearbeitet.

Innovatives Konzept für eine große Geschichte: Jeder Raum des Mercedes-Benz Museums erzählt eine Geschichte und ist gezielt auf einzelne Exponate ausgerichtet. Die Exponate sind die Stars und stehen im Zentrum des Interesses. Das Mercedes-Benz Museum entwickelt sich aus dem Inneren, aus den Inhalten. Inhalt und Hülle sind gleichwertig, ergänzen sich selbstverständlich und erscheinen als ein großes Ganzes.

Ausstellen und Archivieren sind Wesensmerkmale eines Museums, die Zweiteilung in Mythos- und Collections-Räume folgt diesem Ansatz: Die Mythos-Räume exponieren einzelne herausragende Fahrzeuge aus der 120-jährigen Produktgeschichte von Mercedes-Benz, die Collectionen sind Schaudepots, die der reichhaltigen Sammlung mit einer Vielzahl von unterschiedlichen Nutzfahrzeugen und Personenwagen Platz bieten.

HG Merz not only assumed responsibility for the presentation of the collection and the corporate history in the Mercedes-Benz Museum; he was also involved in the development of the concept right from the start.

An innovative concept for a grand history: Each room in the Mercedes-Benz Museum tells a story and specifically focuses on individual exhibits. The latter are the stars and take center stage. The Mercedes-Benz Museum develops from the inside out, with designs strongly based on content. Content and shroud are on a par, complement each other in a natural way and together appear as an impressive entity.

Inherent features of a museum are the activities of displaying and archiving, and this is also the approach chosen for the division into Legend and Collection rooms. The Legend rooms display individual, outstanding vehicles from 120 years of Mercedes-Benz product history, whereas the Collections resemble show depots which accommodate the exhaustive collection with a large number of different commercial vehicles and passenger cars.

DIE MYTHOS-RÄUME
[THE LEGEND ROOMS

Sieben verschiedenen Bilder, die in den Mythos-Räumen präsentiert werden, lassen eine Legende erlebbar werden. In einer Zeitreise erzählen sie die Geschichte der Marke Mercedes-Benz von den Anfängen der Automobilerfindung bis zur globalen Expansion der Marke in der Gegenwart. Sie zeigen das Ineinander von Innovation und Tradition, sie stellen die Herausforderungen der Zeitgeschichte und die entsprechende Reaktion von Mercedes-Benz dar.

Seven different scenarios presented in the Legend rooms provide a tangible experience of a legend. In a journey through time, they tell the story of the Mercedes-Benz brand from the beginnings of automotive invention through to the brand's global expansion at the present time. They demonstrate the symbiosis of innovation and tradition and outline both the challenges of contemporary history and the relevant responses of Mercedes-Benz.

Mythos 1 Pioniere *[Pioneers*
Die Erfindung des Automobils, 1886 bis 1900
The Invention of the Automobile, 1886 - 1900

Mythos 2 Mercedes
Die Geburt der Marke, 1900 bis 1914
Birth of the Brand, 1900 - 1914

Mythos 3 Umbrüche *[Times of Change*
Diesel und Kompressor, 1914 bis 1945
Diesel and Supercharger, 1914 - 1945

Mythos 4 Wunderjahre *[Post-war Miracle*
Form und Vielfalt, 1945 bis 1960
Form and Diversity, 1945 - 1960

Mythos 5 Vordenker *[Visionaries*
Sicherheit und Umwelt, 1960 bis 1982
Safety and the Environment, 1960 - 1982

Mythos 6 Weltbewegend *[Moving the World*
Global und individuell, 1982 bis heute
Global and Individual, 1982 until today

Mythos 7 Silberpfeile *[Silver Arrows*
Rennen und Rekorde
Races and Records

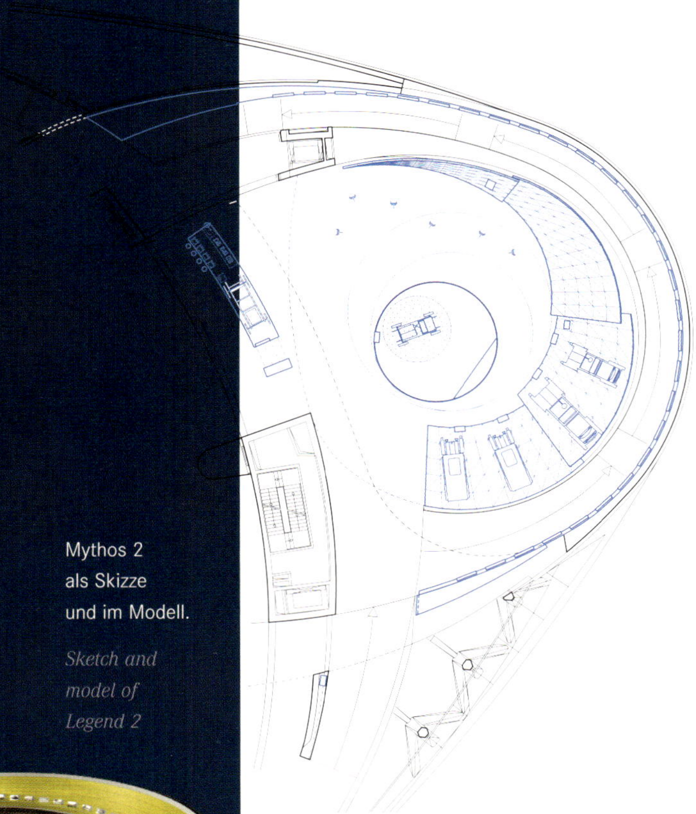

Mythos 2
als Skizze
und im Modell.

*Sketch and
model of
Legend 2*

Die Mythos-Räume setzen sich aus drei Elementen zusammen: Illustrierte Chronik, Szene und Werkbank. Die Illustrierte Chronik gewährt den Besuchern eine generelle Orientierung zu den großen Themen der Zeit- und Unternehmensgeschichte, sensibilisiert sie für das Bild auf der Ebene. Auf der sogenannten Szene, der Ausstellungsfläche im Zentrum jedes Mythos-Raumes, werden die wichtigsten Fahrzeuge und Produktinnovationen der Marke Mercedes-Benz aus einer bestimmten Epoche vorgestellt. Die Werkbänke schließlich konzentrieren sich auf die Vermittlung der wichtigsten technischen und gestalterischen Innovationen.

The Legend rooms consist of three elements: Illustrated History, Scene and Workbench. The Illustrated History provides visitors with information about the major topics of contemporary and corporate history and tunes them in to the scene at the respective level. On the exhibition platform – the so-called Scene – in the center of each Legend room, the most important vehicles and product innovations of the Mercedes-Benz brand from a specific era are displayed. The Workbenches, finally, convey information about the most important technical and design innovations.

◄ **MYTHOS 1**
[LEGEND 1

Pioniere – Die Erfindung des Automobils, 1886 bis 1900
Pioneers – The Invention of the Automobile, 1886 – 1900

MYTHOS 2 ►
[LEGEND 2

Mercedes – Die Geburt der Marke, 1900 bis 1914
Mercedes – Birth of the Brand, 1900 – 1914

MYTHOS 3
[LEGEND 3

Umbrüche –
Diesel und Kompressor,
1914 bis 1945
Times of Change –
Diesel and Supercharger,
1914 – 1945

MYTHOS 4
[LEGEND 4

Wunderjahre – Form
und Vielfalt,
1945 bis 1960

Post war Miracle –
Form and Diversity,
1945 – 1960

1960
–1982
Vordenker

MYTHOS 5
[LEGEND 5

Vordenker – Sicherheit
und Umwelt,
1960 bis 1982
*Visionaries – Safety
and the Environment,
1960 – 1982*

MYTHOS 6
[LEGEND 6

Weltbewegend – Global
und individuell, ab 1982
*Moving the World –
Global and Individual,
from 1982*

DIE COLLECTIONSRÄUME

[THE COLLECTION ROOMS

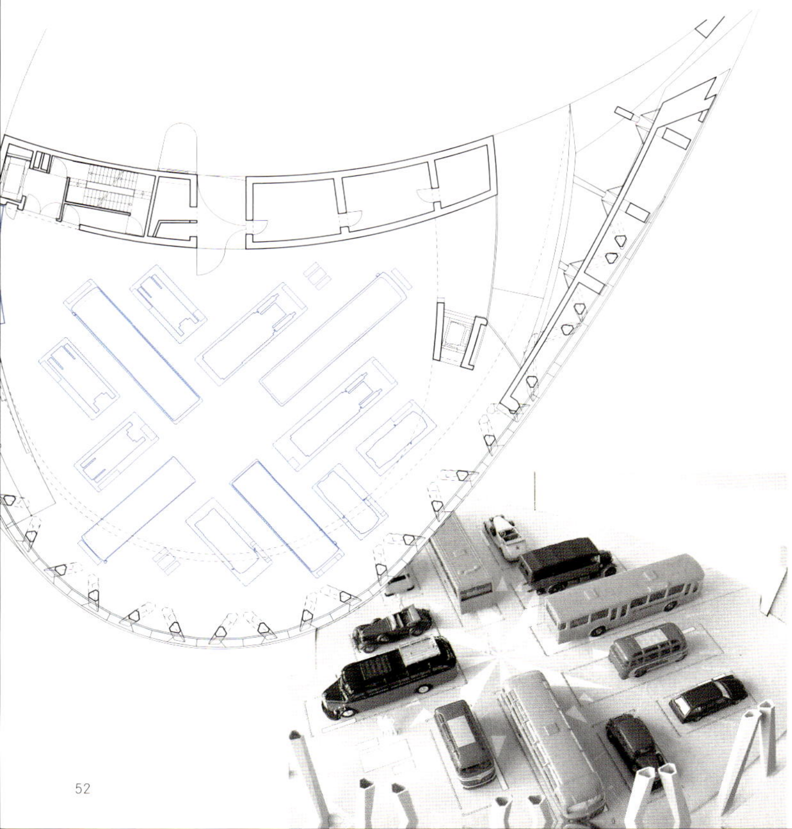

Ein Collectionsraum in Skizze und im Modell.

Sketch and model of a Collection room

COLLECTION 1	Galerie der Reisen *[Gallery of Voyager*
COLLECTION 2	Galerie der Lasten *[Gallery of Carriers*
COLLECTION 3	Galerie der Helfer *[Gallery of Helpers*
COLLECTION 4	Galerie der Namen *[Gallery of Celebrit*
COLLECTION 5	Galerie der Helden *[Gallery of Heroes*

Fünf thematisch geordnete Collectionsräume zeigen Fülle und Vielfalt der Fahrzeuge von Mercedes-Benz. Sie präsentieren die Identität der Marke und ihrer Produkte. Während die Mythos-Räume inszenieren, dokumentieren die Collectionen die Faszination Mercedes-Benz. Mit ihrem Exponatreichtum sind die Collectionen Spiegel von Erfahrung, Kompetenz und Perfektion in der gesamten Breite aller automobilen Produkte.

Die depotartigen Collectionen, angelehnt an thematische Schausammlungen der klassischen Museen, zeigen in einer sachlichen Umgebung eine größtmögliche Fülle von Fahrzeugen und ergänzenden Schaustücken. Die Zusammenstellung und die Zuordnung der einzelnen Exponate ergeben eine „Wunderkammer" des jeweiligen Themas.

Five thematically arranged Collection rooms display the abundance and diversity of Mercedes-Benz vehicles. They present the identity of the brand and its products. Whereas the Legend rooms stage-set the exhibits, the Collections document the fascination of Mercedes-Benz. With their abundance of exhibits, the Collection rooms reflect experience, expertise and perfection across the entire bandwidth of all automotive products.

The Collection rooms, resembling depots in the style of thematic collections of classic museums, display the greatest possible abundance of vehicles and complementary exhibits in a down-to-earth atmosphere. The composition and allocation of the individual exhibits create treasure troves for the individual topics.

56
OTTO WETZ

COLLECTION 1

Galerie der Reisen

Gallery of Voyagers

COLLECTION 2

Galerie der Lasten

Gallery of Carriers

COLLECTION 3

Galerie der Helfer

Gallery of Helpers

Helfen, Retten und Bergen ist oft ein Wettlauf mit der Zeit. Mercedes-Benz Personenwagen und Nutzfahrzeuge sind durch ihre Zuverlässigkeit und Langlebigkeit von Beginn an ideale Fahrzeuge für diese Einsatzzwecke. Seit vielen Jahren bietet Mercedes-Benz Sonderfahrzeuge für spezielle Anforderungen an.

More often than not, every minute counts in help, rescue and salvage operations. Right from the start, Mercedes-Benz passenger cars and commercial vehicles have been ideal for these operations, thanks to their reliability and longevity. Mercedes-Benz has been offering purpose-built vehicles for specific requirements for many years.

GROSS
ADAC
POLIZEI

COLLECTION 4

Galerie der Namen

Gallery of Celebrities

Mercedes-Benz stand schon immer für außergewöhnliche Automobile, die die anspruchsvolle Persönlichkeit des Besitzers unterstrichen. Bis heute vertrauen Könige, Kanzler und Präsidenten, Päpste und Filmstars, Persönlichkeiten aus Sport und Showbusiness den Fahrzeugen der Marke mit dem Stern.

Mercedes-Benz has always been creating out of the ordinary automobiles which underscore the demanding personalities of their owners. To this very day, kings, chancellors and presidents, popes and movie stars as well as personalities from the worlds of sports and showbiz have been relying on cars with the three-pointed star.

COLLECTION 5

Galerie der Helden

Gallery of Heroes

RENNSPORT
[RACING

Ein Höhepunkt der Ausstellung und weltweit einzigartig ist die Ebene „Silberpfeile – Rennen und Rekorde“, die auf einer Steilwandkurve und einer Ausstellungsfläche von rund 1.200 Quadratmetern 40 Highlights aus der langen, traditionsreichen Motorsportgeschichte von Mercedes-Benz zeigt. Vom Daimler „Phönix“-Rennwagen und dem berühmten „Blitzen-Benz“ über die legendären Silberpfeile der 30er und 50er Jahre bis hin zu den siegreichen Formel-1- und DTM-Rennwagen aus dem aktuellen Renngeschehen vermittelt die Ausstellung hautnah die Faszination Motorsport mit seiner mitreißenden Atmosphäre und den einmaligen Erfolgen.

The exhibition culminates in a globally unique scenario named "Silver Arrows – Races and Records". Here, 40 highlights from the long-standing tradition of Mercedes-Benz motor sport are displayed, occupying a steep-bank curve and 1,200 square meters of exhibition space. From the Daimler Phoenix racing car and the famous Lightning Benz via the legendary Silver Arrows of the 1930s and 1950s through to the victorious Formula One and DTM cars entered in present-day competitions, the exhibition gives a vivid impression of the fascination of motor sport with its intoxicating atmosphere and unique successes.

MERING
HEIMRING
BROOKLANDS

„Silberpfeile – Rennen & Rekorde“: Eine Steilwandkurve vermittelt den Besuchern die komplette Motorsportgeschichte von Mercedes-Benz – von den Anfängen um 1900, über die historischen Silberpfeile bin hin zu den Rennserien der Gegenwart.

“Silver Arrows – Races and Records”: A steep-bank curve gives visitors an overview of Mercedes-Benz motor sport history – from the very beginnings around 1900 via the historical Silver Arrows through to the present day race series.

Motorsport 1952 – 1955
Motorsport 1952 – 1955

Count W. Elliott Zborowski Henri Degrais
Walter de Jochems Edgar Ladenburg
Franz Heim Bob Burman Erwin Bergdoll
Richard Schultze-Steprath
–1955
1955
Rallye-Sport
Rally sport

Rennvitrinen mit Ausstellungsstücken aus den jeweiligen Epochen des Motorsports und interaktive Datenbanken lassen bei den Besuchern fast den Eindruck entstehen, man könnte das Grollen der Motoren, den Jubel des Publikums oder die spannenden Positionskämpfe live miterleben.

Showcases with exhibits from the different motor sport eras and interactive databases contribute to the impression on the part of the visitor of being a live witness to the roar of the engines, the cheering of the crowds or the thrilling duels on the racetrack.

REKORDE
[RECORDS

Als Abschluss geht die Steilwandkurve in eine senkrechte Wand über, an der berühmte Rekordfahrzeuge von Mercedes-Benz montiert sind. Die Auswahl reicht vom dreiachsigen T 80 und dem Zwölfzylinder-Rekordwagen W 125 von 1938 über den Dieselrekordler C 111 III von 1978 und den 190 E 2.3-16 von 1983 bis hin zum aktuellen E 320 CDI, der 2005 im texanischen Laredo 100.000 Meilen mit einer Durchschnittsgeschwindigkeit von 224 km/h absolviert hat. Dass Geschwindigkeit aber nicht alles ist, demonstrieren die beiden abschließenden Exponate: Das sogenannte Sparmobil schaffte es, mit nur einem Liter Diesel hochgerechnet 1028 Kilometer zurückzulegen. Ganz ohne herkömmlichen Kraftstoff kam schließlich das von Solarzellen gespeiste Solarmobil aus, das Mitte der 1980er Jahre die Tour del Sol vom Bodensee zum Genfer See gewann.

The steep-bank curve gradually makes way for a vertical wall to which famous record-breaking cars from Mercedes-Benz are fixed, ranging from the three-axle T 80 and the W 125 twelve-cylinder record-breaker from 1938 via the diesel-engined C 111 III record-breaking car of 1978 and the 190 E 2.3-16 from 1983 through to the up-to-date E 320 CDI which covered 100,000 miles at an average speed of 224 km/h (140 mph) in Laredo, Texas, in 2005. But speed is not everything – as demonstrated by the last two exhibits. The so-called economobile covered the equivalent of 3,020 miles per gallon of diesel fuel. And finally, the solar-mobile, powered by solar cells and doing completely without conventional fuel, won the Tour del Sol from Lake Constance to Lake Geneva in the mid-1980s.

Zu den spektakulärsten Fahrzeugen der Mercedes-Benz Geschichte gehören mit Sicherheit die Rekordwagen. Im Mercedes-Benz Museum hängen sie vertikal an der Wand und bilden den Abschluss des Rundgangs.

The record-breaking cars clearly rank among the most spectacular vehicles in Mercedes-Benz history. In the Mercedes Benz Museum, they are attached to the walls and represent the concluding highlights of the tour.

FASZINATION TECHNIK

[THE FASCINATION OF TECHNOLOGY

Die „Faszination Technik“ ist eine eigenständige, für alle Besucher frei zugängige Ausstellung. Modernste Medien und Exponate ermöglichen ein Blick in den Arbeitsalltag der Forscher, Designer, Entwickler und Ingenieure von Mercedes-Benz und damit auch einen Ausblick in die Zukunft des Automobils.

"The Fascination of Technology" is a separate exhibition to which all visitors have free-of-charge access. The most advanced media and exhibits provide a view of the everyday work of researchers, designers, developers and engineers at Mercedes Benz – and thus a glimpse into the future of the automobile.

Die Faszination Technik gewährt einen Einblick in die komplexen Prozesse der Fahrzeugentstehung und eröffnet den Besuchern so die faszinierende Welt „hinter dem Stern“.

“The Fascination of Technology” gives an insight into the complex vehicle development processes and allows visitor to cast a glance at the intriguing world “behind the star”.

DATEN UND FAKTEN
[FACTS AND FIGURES

Ausgestellte Fahrzeuge *[No. of vehicles on display* 160
_ Personenwagen *[Passenger cars* 80
_ Nutzfahrzeuge *[Commercial vehicles* 40
_ Renn- und Rekordfahrzeuge *[Racing and record cars* 40
Kleinexponate *[Small exhibits* 1.300

LED-Ring im Mythos 6 *[LED ring in Legend room no. 6*
_ Oberfläche *[Surface* 28 Quadratmeter *[square meters*
_ Durchmesser *[Diameter* 30 Meter *[meters*

Anzahl Audio-Guides *[No. of audio guides* 2.000 Stück
Sprachen Audio-Guides *[No. of audio guide languages* 8 Sprachen
Gesprochener Audio-Text *[Audio text* 2.500 Minuten (42 Stunden) *[minutes (42 hours)*
Anzahl Audio-Videospieler *[No. of audio/video players* 259 Stück
Anzahl Monitore *[No. of monitors* 235 Stück
Anzahl Media-Kontrolleinheiten *[No. of media control units* 22 Stück
Anzahl Lautsprecher *[No. of speakers* 175 Stück
Anzahl Media-Installationen *[No. of media installations* 90 Stück
Speicherkapazität aller Geräte *[Storage capacity of all units* 25,5 Terabytes
Recherche: Videokassetten/Filme *[Research in video/film footage* 1.800 Stück
Speicher für Filmmaterial *[Storage capacity for footage* 26 Terabytes
Eingesehene Foto-/Filmarchive *[Photo/film archives searched* 47 Stück
Zurückgelegte Flugstrecke für Filme *[Air miles covered for film shooting* 422.500 Kilometer
Produzierte Filme *[Length of films produced* 500 Minuten *[minutes*
Fotos *[Photos* 358 Stück
Überschriften *[Headlines* 709 Stück
Wörter *[Printed words* 89.000 Stück

Über 1.500 Kleinexponate finden die Besucher in den Schaukästen der Collectionsräume.

The showcases in the Collection rooms display over 1,500 small items.

DIE EXPONATE: PERSONENWAGEN
[THE EXHIBITS: PASSENGER CARS

BAUJAHR | TYP | STANDORT IM MUSEUM
[YEAR OF CONSTRUCTION | MODEL | LOCATION IN THE MUSEUM

1885	Daimler Motorkutsche	MYTHOS 1
1886	Daimler Einzylindermotor „Standuhr“	MYTHOS 1
1886	Benz Patent-Motorwagen	MYTHOS 1
1886	Daimler Motorkutsche	MYTHOS 1
1889	Daimler Motor-Quadricycle „Stahlradwagen“	MYTHOS 1
1892	Daimler Motor-Straßenwagen	COLLECTION 4
1893	Benz Vis-à-Vis	MYTHOS 1
1894	Benz Motor-Velociped	MYTHOS 1
1896	Daimler „Riemenwagen“ Vis-à-Vis	MYTHOS 1
1899	Benz Dos-à-Dos	MYTHOS 1
1902	Mercedes-Simplex 40 PS	MYTHOS 2
1904	Mercedes-Simplex 60 PS Reiselimousine	MYTHOS 2
1905	Benz 18 PS Doppelphaeton	MYTHOS 2
1908	Mercedes 75 PS Doppelphaeton	MYTHOS 2
1909	Benz 20/35 PS Landaulet	MYTHOS 2
1921	Mercedes-Knight 16/45 PS Tourenwagen	COLLECTION 1

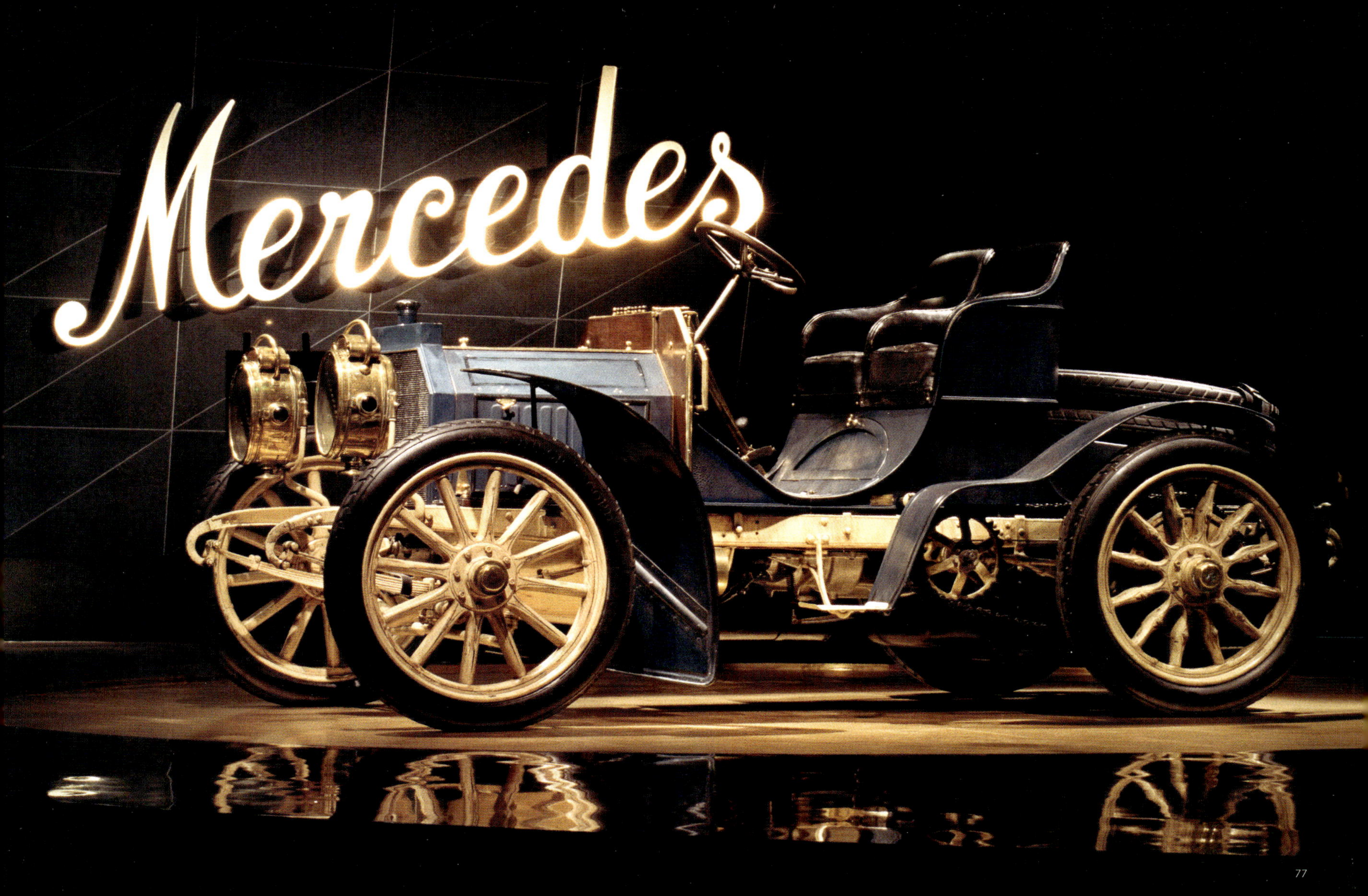
Mercedes

1923	Mercedes 10/40 PS Sport-Zweisitzer	MYTHOS 3
1923	Benz 10/30 PS	COLLECTION 5
1926	Mercedes-Benz 24/100/140 PS Roadster	COLLECTION 4
1927	Mercedes-Benz 12/55 PS Pullmann-Limousine	COLLECTION 1
1927	Mercedes-Benz 8/38 PS	COLLECTION 5
1928	Mercedes-Benz 26/120/180 PS Typ S Tourenwagen	MYTHOS 3
1928	Mercedes-Benz 27/170/225 PS Typ SSK Sport-Zweisitzer	MYTHOS 3
1931	Mercedes-Benz 170	COLLECTION 5
1932	Mercedes-Benz 770 „Großer Mercedes“ Cabriolet F	COLLECTION 4
1935	Mercedes-Benz 130	COLLECTION 5
1935	Mercedes-Benz 770 „Großer Mercedes“ Pullmann-Limousine	COLLECTION 4
1936	Mercedes-Benz 500 K Spezial-Roadster	MYTHOS 3
1937	Mercedes-Benz 540 K Cabriolet B	MYTHOS 3
1937	Mercedes-Benz 770 „Großer Mercedes“ offener Tourenwagen	MYTHOS 3
1937	Mercedes-Benz 320 Krankenwagen	COLLECTION 3
1938	Mercedes-Benz 260 D Pullmann-Limousine	MYTHOS 3
1939	Mercedes-Benz 320 Stromlinien-Limousine	COLLECTION 1
1951	Mercedes-Benz 170 D	COLLECTION 5
1952	Mercedes-Benz 300	MYTHOS 4
1954	Mercedes-Benz 300 S Cabriolet A	MYTHOS 4
1955	Mercedes-Benz 180	MYTHOS 4
1955	Mercedes-Benz 300 SL Coupé	MYTHOS 4
1955	Mercedes-Benz 300 SLR „Uhlenhaut-Coupé“	MYTHOS 4
1958	Mercedes-Benz 190 SL	COLLECTION 4
1959	Mercedes-Benz 300	COLLECTION 4
1960	Mercedes-Benz 300 Messwagen	MYTHOS 5
1962	Mercedes-Benz 190	COLLECTION 5
1962	Mercedes-Benz 300 SL Roadster	MYTHOS 4

1964	Mercedes-Benz 220 S	MYTHOS 5
1964	Mercedes-Benz 230 SL „Pagode“	MYTHOS 5
1965	Mercedes-Benz 600 Pullmann Staatslimousine	COLLECTION 4
1969	Mercedes-Benz 250 CE	COLLECTION 5
1972	Mercedes-Benz 300 SEL 6.3	COLLECTION 1
1973	Mercedes-Benz Experimentier-Sicherheits-Fahrzeug ESF22	MYTHOS 5
1975	Mercedes-Benz 240 D Taxi	COLLECTION 5
1975	Mercedes-Benz 200 D Bestattungswagen	COLLECTION 5
1980	Mercedes-Benz 300 SD	MYTHOS 5
1980	Mercedes-Benz 230 G „Papamobil“	COLLECTION 4
1982	Mercedes-Benz 240 D	COLLECTION 5
1983	Mercedes-Benz 500 SEC	COLLECTION 2
1984	Mercedes-Benz 190 E	MYTHOS 5
1984	Mercedes-Benz 190 E 2.3 AMG	COLLECTION 4
1985	Mercedes-Benz 300 TD	COLLECTION 1
1985	Mercedes-Benz 230 E	COLLECTION 2
1985	Mercedes-Benz 280 CE	COLLECTION 2
1985	Mercedes-Benz 380 SEL	COLLECTION 2
1988	Mercedes-Benz 200 D	MYTHOS 6
1988	Mercedes-Benz 500 SL	COLLECTION 4
1991	Mercedes-Benz 500 SL	COLLECTION 4
1995	Mercedes-Benz S 600	MYTHOS 6
1996	Mercedes-Benz E 200 T-Modell	COLLECTION 5
1997	Mercedes-Benz ML 320	COLLECTION 4
1998	Mercedes-Benz 290 GD Turbodiesel	MYTHOS 6
2000	Mercedes-Benz C220 CDI T-Modell Polizei	COLLECTION 3
2001	Mercedes-Benz C32 AMG T-Modell F 1 Medical Car	COLLECTION 3

DIE EXPONATE: NUTZFAHRZEUGE
[THE EXHIBITS: COMMERCIAL VEHICLES

BAUJAHR [YEAR OF CONSTRUCTION	TYP [MODEL	STANDORT IM MUSEUM [LOCATION IN THE MUSEUM
1892	Daimler Motor-Feuerspritze	MYTHOS 1
1895	Benz Omnibus	MYTHOS 1
1898	Daimler Motor-Lastwagen	MYTHOS 1
1899	Daimler Motor-Geschäftswagen	MYTHOS 1
1907	Milnes-Daimler Doppeldeckerbus	COLLECTION 1
1912	Benz 3-Tonnen-Lastwagen	COLLECTION 2
1912	Benz Feuerwehr-Motorspritze	COLLECTION 3
1928	Mercedes-Benz Dieselschlepper Typ OE	COLLECTION 5
1932	Mercedes-Benz Lo 2000 Diesel Pritschenwagen	MYTHOS 3
1937	Mercedes-Benz L 1500 Pritschenwagen mit Holzvergaser	COLLECTION 5
1938	Mercedes-Benz L 6500 Pritschenwagen	COLLECTION 2
1938	Mercedes-Benz O 10000 mobiles Postamt	COLLECTION 2
1940	Mercedes-Benz O 2600 Allwetter-Reisebus	COLLECTION 1
1949	Unimog 25 PS	COLLECTION 5
1952	Mercedes-Benz O 3500 Allwetterreisebus	COLLECTION 1

1952	Mercedes-Benz LF 3500 Feuerwehrfahrzeug mit Metz Drehleiter DL 22	COLLECTION 3
1952	Mercedes-Benz 170 V Kastenwagen	COLLECTION 3
1955	Mercedes-Benz Rennwagen-Schnelltransporter	COLLECTION 2
1959	Mercedes-Benz LP 333 Pritschenwagen	COLLECTION 2
1960	Mercedes-Benz LK 338 Kipper	MYTHOS 4
1965	Mercedes-Benz L 406 Kastenwagen	COLLECTION 2
1969	Mercedes-Benz LO 1112 Omnibus	COLLECTION 1
1974	Mercedes-Benz LP 1513 Heizöl-Tankwagen	COLLECTION 2
1974	Mercedes-Benz O 302 Mannschafts-Reisebus	COLLECTION 4
1979	Mercedes-Benz O 303 Reise-Omnibus	MYTHOS 5
1979	Mercedes-Benz L 409 Pritschenwagen mit Doppelkabine	COLLECTION 3
1980	Mercedes-Benz 1624 Autotransporter	COLLECTION 2
1991	Mercedes-Benz 208 D Verkaufswagen	COLLECTION 5
1992	Mercedes-Benz 814 Bergungsfahrzeug	COLLECTION 3
1994	Mercedes-Benz 508 D Kastenwagen	COLLECTION 2
1999	Mercedes-Benz Sprinter 312 D mobile Zahnarztpraxis	MYTHOS 6
2001	Mercedes-Benz Sprinter 313 CDI Rettungswagen	COLLECTION 3
2003	Mercedes-Benz Actros 3341 A	MYTHOS 6
2004	Mercedes-Benz Unimog U 500 Winterdienstfahrzeug	COLLECTION 3
2005	Mercedes-Benz Viano MARCO POLO CDI 2.2	COLLECTION 1
2005	Mercedes-Benz Travego Reisebus	COLLECTION 1
2005	Mercedes-Benz Econic 2628 NGT Müllsammelfahrzeug	COLLECTION 3

DIE EXPONATE: RENNFAHRZEUGE
[THE EXHIBITS: RACING CARS

BAUJAHR | TYP | STANDORT IM MUSEUM
[YEAR OF CONSTRUCTION | MODEL | LOCATION IN THE MUSEUM

1900	Benz 14 PS Rennwagen	RENNEN & REKORDE
1900	Daimler „Phönix“ 23 PS Rennwagen	RENNEN & REKORDE
1908	Benz Grand-Prix-Rennwagen	RENNEN & REKORDE
1914	Mercedes Grand-Prix-Rennwagen	RENNEN & REKORDE
1924	Mercedes 2-Liter-Rennwagen „Targa Florio“	RENNEN & REKORDE
1934	Mercedes-Benz 750-kg-Rennwagen W 25	RENNEN & REKORDE
1937	Mercedes-Benz Stromlinienrennwagen W 25 „Avus“	RENNEN & REKORDE
1937	Mercedes-Benz 750-kg-Rennwagen W 125	RENNEN & REKORDE
1938	Mercedes-Benz 3-Liter-Rennwagen W 154	RENNEN & REKORDE
1939	Mercedes-Benz 1,5-Liter-Rennwagen W 165 „Tripolis“	RENNEN & REKORDE
1939	Mercedes-Benz 230 S Geländesportwagen	RENNEN & REKORDE
1952	Mercedes-Benz 300 SL Rennsportwagen	RENNEN & REKORDE
1955	Mercedes-Benz 2,5-Liter-Stromlinien-Rennwagen W 196 R	RENNEN & REKORDE
1955	Mercedes-Benz 2,5-Liter-Rennwagen W 196 R	RENNEN & REKORDE
1955	Mercedes-Benz 300 SLR Rennsportwagen	RENNEN & REKORDE
1955	Mercedes-Benz 300 SLR Rennsportwagen	COLLECTION 2

1963	Mercedes-Benz 300 SE Ralleywagen	RENNEN & REKORDE
1977	Mercedes-Benz 280 E Ralleywagen	RENNEN & REKORDE
1980	Mercedes-Benz 500 SLC Ralleywagen	RENNEN & REKORDE
1989	Sauber-Mercedes Gruppe-C-Rennsportwagen C 9	RENNEN & REKORDE
1990	Sauber-Mercedes Gruppe-C-Rennsportwagen C 11	RENNEN & REKORDE
1990	Mercedes-Benz 1450 S Renntruck	RENNEN & REKORDE
1992	AMG-Mercedes 190 E 2.5-16 Evolution II DTM-Tourenwagen	RENNEN & REKORDE
1994	Penske-Mercedes IndyCar PC 23	RENNEN & REKORDE
1995	AMG-Mercedes C-Klasse DTM-Tourenwagen	RENNEN & REKORDE
1997	Mercedes-Benz GT-Rennsportwagen CLK-GTR	RENNEN & REKORDE
1997	McLaren-Mercedes Formel-1-Rennwagen MP 4-12	RENNEN & REKORDE
1998	Mercedes-Benz Atego Renntruck	RENNEN & REKORDE
1998	McLaren-Mercedes Formel-1-Rennwagen MP 4-13	RENNEN & REKORDE
1999	McLaren-Mercedes Formel-1-Rennwagen MP 4-14	RENNEN & REKORDE
2001	AMG-Mercedes CLK DTM-Tourenwagen	RENNEN & REKORDE
2005	AMG-Mercedes C-Klasse DTM-Tourenwagen	RENNEN & REKORDE

DIE EXPONATE: REKORDFAHRZEUGE
[THE EXHIBITS: RECORD CARS

BAUJAHR | TYP | STANDORT IM MUSEUM
[YEAR OF CONSTRUCTION | MODEL | LOCATION IN THE MUSEUM

1938	Mercedes-Benz Zwölfzylinder-Rekordwagen W 125	RENNEN & REKORDE
1939	Mercedes-Benz Weltrekordwagen T 80	RENNEN & REKORDE
1978	Mercedes-Benz C 111-III Diesel-Rekordwagen	RENNEN & REKORDE
1980	Mercedes-Benz Kraftstoff-Sparmobil	RENNEN & REKORDE
1983	Mercedes-Benz 190 E 2.3-16 Rekordwagen	RENNEN & REKORDE
1985	Mercedes-Benz / Alpha real Solarmobil	RENNEN & REKORDE
2005	Mercedes-Benz E 320 CDI Rekordwagen	RENNEN & REKORDE

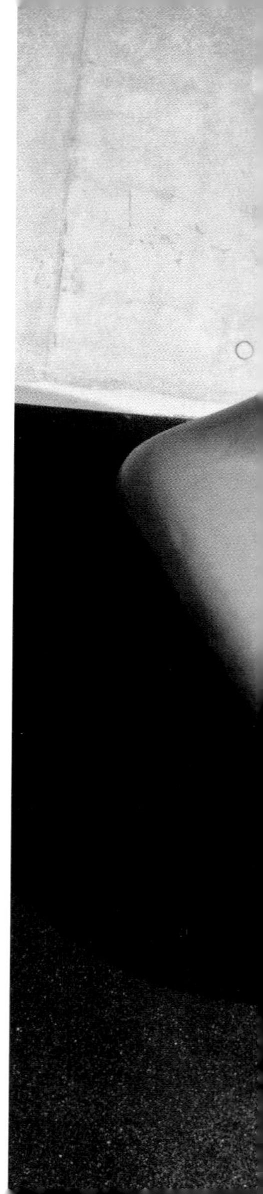

DIE PASSAGE [THE PASSAGE

In der 80 Meter langen Passage können sich die Besucher bequem zwischen dem Mercedes-Benz Museum und dem Mercedes-Benz Center Stuttgart bewegen. Die Passage stellt nicht nur die räumliche Verbindung zwischen der Tradition und der Gegenwart dar, sondern bildet eine eigene, attraktive Erlebniswelt. Das Angebot reicht von abwechslungsreicher Gastronomie über Shops mit hochwertigen Produkten aus den Mercedes-Benz Collectionen bis hin zu Serviceangeboten wie einem Geldautomaten, einem Briefkasten oder einer öffentlichen Telefonzelle.

In the 80-meter long passage, visitors are able to move between the Mercedes-Benz Museum and the Mercedes-Benz Center Stuttgart in all comfort. The passage not only physically connects tradition and present but also forms an attractive world of experience in its own right. Services range from different restaurants to shops offering high grade products from the Mercedes-Benz Collections, from cash terminal and mail box to phone booth.

Offene, lichtdurchflutete Shops laden die Besucher zum Verweilen ein, egal ob vor oder nach dem Museumsbesuch.

Open, light-flooded shops invite visitors to linger, before or after a visit to the museum.

ADRESSE UND ÖFFNUNGSZEITEN
[ADDRESS AND OPENING HOURS

ANSCHRIFT [ADDRESS

Mercedes-Benz Museum
Mercedesstraße 100
70372 Stuttgart

ÖFFNUNGSZEITEN [OPENING HOURS

Di bis So & an Feiertagen
von 9.00 bis 18.00 Uhr
Montags geschlossen
Die Kasse schließt um 17.00 Uhr

Tuesday - Sunday and on public holidays
from 9 a.m. until 6 p.m.
Closed on Mondays
The ticket counter closes at 5 p.m.

Die Restaurants in der Passage gliedern sich nahtlos in die einzigartige Architektur des Mercedes-Benz Museums ein.

The restaurants in the passage are harmoniously integrated in the unique architecture of the Mercedes-Benz Museum.

DIE RESTAURANTS

[THE RESTAURANTS

Die Gastronomie in der Passage ist – wie das Mercedes-Benz Museum auch – einzigartig: Sechs unterschiedlich gestaltete Restaurantinseln mit fast 300 Plätzen laden zum Schlemmen und Genießen ein. Von schwäbischen Spezialitäten bis hin zu wöchentlich wechselnden Angeboten, die Speisekarte der Restaurants lässt keine Wünsche offen. Auch für Abendveranstaltungen kann die Gastronomie des Mercedes-Benz Museums gebucht werden.

Like the Mercedes-Benz Museum itself, the catering in the passage is quite unique. Six different restaurant islands with capacity for almost 300 people invite visitors to wine and dine. The restaurants' menus leave nothing to be desired, offering a wide choice from Swabian specialties to seasonal dishes changing at weekly intervals. The catering areas of the Mercedes-Benz Museum can also be booked for evening events.

IMPRESSUM
[IMPRINT

Konzeption: Max-Gerrit von Pein, Stuttgart
Texte: Enrico Müller
Bildnachweis:
Archivfotos: DaimlerChrysler Konzernarchiv, Stuttgart
UNStudio, Amsterdam: Seiten 17, 27, 28, 29
HG Merz, Stuttgart: Seiten 40, 41, 43, 52
Übersetzung: DaimlerChrysler Language Management (CFM/LM)
Satz und Grafik: Cora von Pein, München
Reproduktion: immedia23, Stuttgart
Gesamtherstellung: Druckerei Uhl, Radolfzell
Verlag und Vertrieb:
Stadler Verlagsgesellschaft, Konstanz
Max-Stromeyer-Straße 172
78467 Konstanz
Homepage: www.verlag-stadler.de
E-Mail: info@verlag-stadler.de

Concept: Max-Gerrit von Pein, Stuttgart
Texts: Enrico Müller
Photos:
Archive photos: DaimlerChrysler Corporate Archives,
UNStudio, Amsterdam: pages 17, 27, 28, 29
HG Merz, Stuttgart: pages 40, 41, 43, 52
Translation: DaimlerChrysler Language Management (CFM/LM)
Typesetting and graphic design: Cora von Pein, München
Reproduction: immedia23, Stuttgart
Printing: Druckerei Uhl, Radolfzell
Publishing and marketing:
Stadler Verlagsgesellschaft, Konstanz
Max-Stromeyer-Straße 172
78467 Konstanz
Home page: www.verlag-stadler.de
E-mail: info@verlag-stadler.de

ISBN-13: 978-3-7977-0532-7
ISBN-10: 3-7977-0532-8

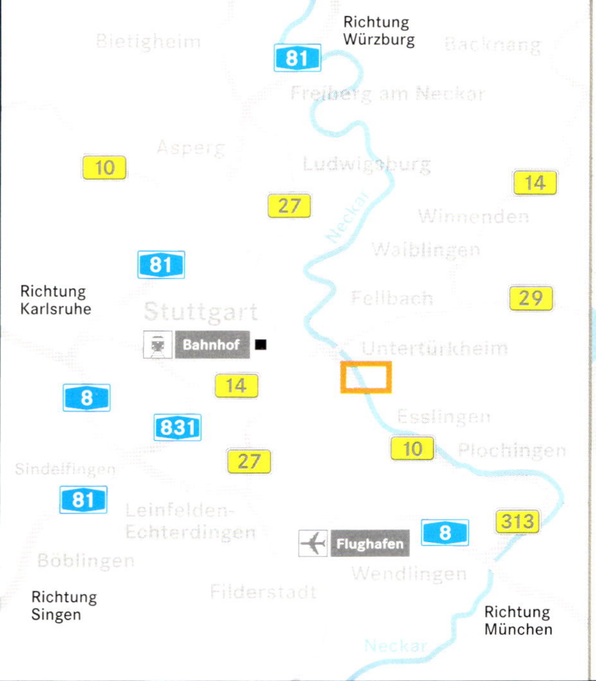

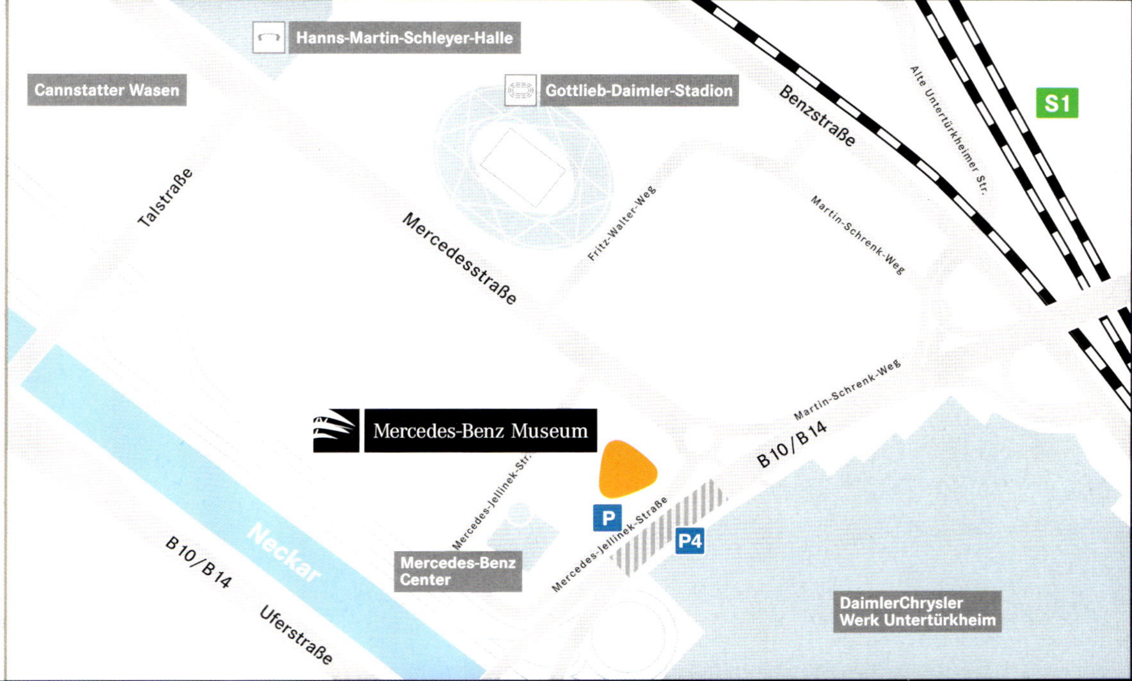

Orientierungshilfen im Straßenverkehr | Ziel | Parken | markante Orientierungspunkte | S-Bahn-Haltestelle | Neckar